TOP-SECRET NATURE

WHERE RAIN COMES FROM

MARIE ROGERS

New York

Published in 2021 by The Rosen Publishing Group, Inc.
29 East 21st Street, New York, NY 10010

First Edition

Editor: Amanda Vink
Book Design: Rachel Rising

Portions of this work were originally authored by Abby Wilson and published as *What Makes Rain?* All new material in this edition authored by Marie Rogers.

Photo Credits: Cover, p.1 peresanz/Shutterstock.com; pp.4,6,8,10,12,14,16,18,20,22 (background) cluckva/Shutterstock.com; p. 5 A3pfamily/Shutterstock.com; p. 7 MNStudio/Shutterstock.com; p. 9 Koldunov Alexey/Shutterstock.com; p. 11 VRstudio/Shutterstock.com; p.13 KC Lens and Footage/Shutterstock.com; p. 15 patpitchaya/Shutterstock.com; p.17 Martchan/Shutterstock.com; p. 19 Batshevs/Shutterstock.com; p. 21 Jon Manjeot/Shutterstock.com (desert); p.21 Taras Vyshnya/Shutterstock.com (rain forest); p. 22 Mikalai Nick Zastsenski/Shutterstock.com.

Cataloging-in-Publication Data

Names: Rogers, Marie.
Title: Where rain comes from / Marie Rogers.
Description: New York : PowerKids Press, 2021. | Series: Top-secret nature | Includes glossary and index.
Identifiers: ISBN 9781725317512 (pbk.) | ISBN 9781725317536 (library bound) | ISBN 9781725317529 (6 pack)
Subjects: LCSH: Rain and rainfall--Juvenile literature.
Classification: LCC QC924.7 R64 2021 | DDC 551.57'7--dc23

Manufactured in the United States of America

CPSIA Compliance Information: Batch #CSPK20. For Further Information contact Rosen Publishing, New York, New York at 1-800-237-9932.

CONTENTS

Rainy Days

Many times, you can't play outside on a rainy day. You might wish it was sunny and warm all the time. But rain is a good thing! Rain gives us water, which is the most important thing on Earth. Water's necessary for life!

Water's Uses

Water on Earth does many things. Rain gives us water to drink and helps our plants grow. It makes lakes and rivers, and it keeps the oceans deep. And when it's hot, water helps cool us off!

The Water Cycle

Have you ever heard of the water **cycle**? The water cycle shows us the different forms water takes. All water has been around since Earth was young. One of the **stages** of the water cycle is rain.

Warming Up

The water in rain comes from our lakes, oceans, and rivers. When the sun shines, it heats the water. The water **evaporates** and takes the form of **vapor**. You can't see vapor, but it rises into the air and forms clouds.

Water Vapor

The air gets cooler the higher up it goes. When it's cool enough, water vapor turns into tiny water droplets. Clouds collect thousands of droplets. As more vapor turns into water droplets, the clouds get bigger and bigger.

Big Clouds

The **temperature** of the air also helps create rain. When warm air and cold air meet, the warm air pushes cold air higher into the sky. The cold air makes clouds bigger. This pushes more and more water droplets together.

A Cloud Is Cold!

It gets very cold inside clouds! There are even little pieces of ice. The ice mixes with water and makes the water droplets big and heavy. The drops keep forming in the clouds until it's time to rain.

It's Raining, It's Pouring

When the clouds can't hold more water droplets, it starts to rain! Rain is a kind of **precipitation**. Sometimes it only rains for a few minutes. Other times, it rains all day. Has it rained near you lately?

clouds
rain
water vapor
runoff

Water on Earth

All areas of Earth need water. Some places get a lot of rain, and others get very little. It may rain every day in the rain forest. Rain forests are very wet. It doesn't rain a lot in the desert. Deserts are very dry.

desert

rain forest

The Water Cycle Continues

Once it's on the ground, water goes from high places to low places. This moving water is called runoff. The water flows from streams to lakes and even to the oceans. From there, it evaporates and continues the water cycle!

GLOSSARY

cycle: A set of events or actions that repeat.

evaporate: To change from a liquid to a gas.

precipitation: Water that falls to the ground as rain or snow or in other forms.

stage: A step in the growth or change of something.

temperature: How hot or cold something is.

vapor: Matter in the form of a gas or very small drops mixed with air.

INDEX

WEBSITES

Due to the changing nature of Internet links, PowerKids Press has developed an online list of websites related to the subject of this book. This site is updated regularly. Please use this link to access the list: www.powerkidslinks.com/tsn/rain